The Simple Mechanics of the Universe

by Patrick Donlon

DORRANCE PUBLISHING CO
EST. 1920
PITTSBURGH, PENNSYLVANIA 15238

The contents of this work, including, but not limited to, the accuracy of events, people, and places depicted; opinions expressed; permission to use previously published materials included; and any advice given or actions advocated are solely the responsibility of the author, who assumes all liability for said work and indemnifies the publisher against any claims stemming from publication of the work.

All Rights Reserved
Copyright © 2019 by Patrick Donlon

No part of this book may be reproduced or transmitted, downloaded, distributed, reverse engineered, or stored in or introduced into any information storage and retrieval system, in any form or by any means, including photocopying and recording, whether electronic or mechanical, now known or hereinafter invented without permission in writing from the publisher.

Dorrance Publishing Co
585 Alpha Drive
Pittsburgh, PA 15238
Visit our website at www.dorrancebookstore.com

ISBN: 978-1-4809-9167-5
eISBN: 978-1-4809-9381-5

What Makes the World Go Round?

The two main forces produced by the sun are gravity and thrust. Gravity pulls things directly in towards the sun, while thrust pushes things away from the sun. Any object that rotates with speed produces a force of thrust. The thrust of the sun goes out from it in all directions and way out into space. Imagine a woman on a dance floor with a long dress. As she spins on the floor, her dress begins to go out from her body, and as it goes outward, it also curves upward. The woman's' dress follows her body around as she spins, and the dress goes outward. It does not go straight out from her body. The thrust of the sun works the same way. As the thrust of the sun follows the sun around as it rotates, and goes out throughout the solar system and beyond, it sweeps everything in the solar system around the sun. That's what makes planets go around the sun. Gravity pulls objects in towards the sun, while thrust is pushing things away from the sun. And it is where these two forces equalize that a planet is kept on an orbit as it goes around the sun. This is what keeps planets on an orbit at a certain distance from the sun. All day long the thrust of the sun is pushing on the side of the Earth that faces the sun. On the night side of the Earth, there is no thrust. On the front side of the Earth, as it moves forward, the thrust is moving in the same direction. On the back side of the Earth, thrust is pushing the Earth forward on its orbit. The thrust of the sun is striking in the late afternoon and evening which is causing the Earth to rotate into the night where there is no resistance. The Earth is rotating into the east. It is like a basketball player spinning the ball on his finger. They strike across the back side of the ball, causing it to rotate. It is a fact that there is no thrust on the night side of the Earth, which makes it still, compared to the day, which is busy.

How Does Gravity Work?

At the center of the core of every planet, moon, and star, there is a big black hole. A black hole in a star is many times bigger than that of a planet. Fire burns out at the center and draws inward. If you have ever sat in front of a fireplace with the fire burning for a while, you may notice that the center burns out first and draws inward. A flame needs to be fueled and fed. If you took a wet blanket and put it in front of the fireplace, you would suffocate the fire. A star is different than a flame. A star is self-fueling. The Earth has a gravity of one, while the sun, many times bigger, has a gravity of twenty-seven. At the center of the core of the Earth is a big black hole. Just inside the hole is where gravity is produced and then runs back out into space in all directions, way beyond the Moon. Gravity pulls everything from above, back in towards the center of the Earth, coming down through everything on its way. It keeps everything perpendicular to the face of the Earth. Gravity does not hold us down on the Earth by sucking on the soles of our feet, but by coming down through us, like it does to everything else. Gravity pulls the atmosphere in towards the center of the Earth from all directions, which gives us atmospheric pressure that prevails from day to day. The closer you go to the center of the Earth, the higher the pressure produced by gravity. This atmospheric pressure also pressurizes the waters in the oceans. The deeper in the ocean you go, the higher the pressure. The higher above sea level you are, the lower the pressure.

The human body is built for a gravity of one. The solid part of the Earth is like a cast engine; it is porous. Waters deep in the ocean are pressurized in through veins in the Earth that can run for thousands of miles and then appear on hills and mountains as the source of rivers and streams. Rivers and streams don't rise in a valley. Water under pressure will run to the point of least pressure. That is why rivers and streams rise on hills and mountaintops. The higher the altitude, the lower the atmospheric pressure. Most major rivers on the Earth rise up in the north and flow south to the sea. Gravity can turn mud into shale in 150 years or less. People have been picking stones out of the fields for centuries, and they keep on growing from the soil. It is gravity that is producing stones in the soil. Poor soil and a wet climate will produce a lot of stones. There is no stone in the universe made without water. Water is the glue that binds dust or ashes together to form a rock. Living water is the source of life; without it there would be nothing. Gravity shrinks the human body half an inch from the moment you get up in the morning to when you go to bed at night. It is gravity and thrust that hold everything together in the universe and keep order and harmony in the night sky. You may have noticed large stones growing in slow moving streams. These stones grow from dirt and silt in the river beds. You won't find any stones in the middle of the Sahara desert because there is no water to glue the dust together. It's like making a cake. You can put the flour in, but you can't make dough without water. The same is true with concrete or mortar. Nothing was ever created or grew in the universe without water. Water is the glue that binds dust and ashes together to become moons and planets and rocks.

Stars and Galaxies

Planets, stars, and galaxies all rotate into the east. All have a black hole at the center, which is the center of gravity. Galaxies have a black hole, but it is not the same as planets and stars. A hole is something that has to be formed by some material, of which we have many examples. There is no such thing as a hole in the middle of nothing. The hole at the center of a galaxy is formed by the stars at the center of the galaxy, which are very big stars. The stars at the center of a galaxy are many times bigger than the stars at the outskirts of a galaxy, where we are located here in the Milky Way. It is these large stars at the center of a galaxy producing powerful forces of thrust and gravity that form the hole at the center. The thrust of these large stars at the center push each other away, and it is gravity that pulls them back together, which forms the hole. Imagine a group of people on a dance floor forming a circle and dancing around in a ring. As they dance around, they are pushed outward, but with the grips of their hands, they hold the ring together. The fact that a star rotates with speed causes it to move forward, causing the whole galaxy to rotate because all the stars in a galaxy are connected to each other by the forces of gravity and thrust. Gravity between any two stars pulls those two stars towards each other, while thrust from those two stars push each other away. It is where these forces equalize that keeps stars in their position in relation to each other in the night sky. Each star is connected to all the neighboring stars around it and so on, just like reflectors on the spokes of a bicycle wheel. Our sun is moving forward around with the Milky Way at 140 miles per second. Stars closer to the center of the galaxy would be moving a lot slower. Stars produce dust and ashes and throw off clouds of gas. Ashes to ashes and dust to dust. The Earth throws back all the ashes back onto the lunar orbit because ashes are lighter than dust. The dust from the sun is landing on the Earth all day long. God created the Earth over nine thousand years ago, and no ashes every landed on the Earth from the sun. Other planets grew out of dust and ashes on orbits around the sun. It is the thrust of the Earth as it rotates that throws back ashes onto the lunar orbit, where the Moon grew. Galaxies rotate with speed into the east and move forward. They form an oval-shaped track of power and light, like the Earth's orbit around the sun. One end is broad, and the other end is narrow, with an oval-shaped empty field in the middle. Across this field are forces of gravity and thrust holding it together. Galaxies, like the Earth, are tilted on an axis as they rotate. They are tilted back as they move forward, going up on the right side of the track, from the broad end to the narrow end. Just like the Earth, they move backwards coming back down on the left of the track. This is why the Earth moves faster going from winter to summer than it does from summer back down to winter again. It is the fact that galaxies like the Earth move faster going up on the right side that causes them to go farther away from the center, giving us the oval-shaped orbit. This oval-shaped ring of fire was created by God over nine thousand years ago and is called Excelsior, which means ever upward. All life is moving forever upward in the universe. Planets here in our solar system are on orbits of dust and ashes as they go around the sun. An orbit is located where the forces of gravity and thrust equalize as they go out from the sun. The lower down on the sun, the closer the orbit is, and the higher up, the farther out it goes. This is what gives the curvature effect of our solar system. It is not space that is curved; space does not exist. Space is simply the nothingness of the universe. The universe is simply infinite volumes of nothingness that goes on and on forever and ever in all directions without end.

Dating Objects

When measuring and dating objects, we need to take everything that affects an object into account. There are several theories out there that talk about millions of years ago. Man has only lived on this planet for a little over nine thousand years. The Earth is the oldest body in the universe and is the mother of all planets, moons, and rocks in the universe. There is no rock in existence anywhere without water. Living water is the source of life. Water is the glue that binds dust or ashes together to become a rock. Dust or ashes on their own can't form a rock. All the water in the universe came from here, planet Earth. There are things to be taken into account when dating objects, like gravity for example. People have counted new moons growing around Saturn and Jupiter in the last fifty years, which is not very long to grow a rock. The sun has been throwing dust on the Earth now for nine thousand years, which must have added up to something. We can know that the Earth is growing because the moon is moving away from the Earth a half inch per year. Because the Earth is growing, it is producing more thrust. Gravity can grow stones in the field in a very short time. Another thing we can look at is the flood, the biggest event in the history of the planet, seven thousand years ago. During the flood, the sky, which God built two thousand years before, fell back down on the Earth, covering everything in water. The Earth, like today, would have been top heavy, since most of the land mass was in the northern hemisphere. Once submerged under water, any object that is top heavy will turn upside down. In a very short period of time, as the Earth turned upside down, it would have been rotating against the flow of the oceans, which is a powerful force. A lot of damage would have been done. All kinds of vegetation would have been ripped out of the ground, piled up and smothered in mud, some of which has been turned into coal today. Villages and towns would have been leveled, schools of fish buried in mud and would have been fossilized over the years. People in Ireland have dated the bogs of Ireland to be seven thousand years old. Forests of oak trees were uprooted and laid on top of each other but were not covered with mud. Today you can still find hard pieces of oak trees not completely rotted in the bog. A lot of other damage was done by the flood, and evidence of it is all over the face of the Earth and on the ocean floor. There was no Ice Age. During the flood, as the solid part of the Earth turned upside down, the core of the Earth would have taken longer to turn upside down, which would have caused volcanoes to erupt and the Earth to quake, creating canyons and rocks that can only be formed under water. After the flood, all the land mass would have been in the southern hemisphere. After Noah's time, when people grew in numbers again, they started going the wrong way all over again. God saw this and intervened. He scattered the people and confused the tongue, giving each group their own language, and they spread out over the land mass, as is written in the Bible. As the people moved out over the land mass, the continental plates started to shift, pushing east, west and north, back to where we are today. This is how people ended up over all the different continents and islands all over the world. Facts are easy to explain when you take into account what affects them.

-

The Hole in the Ozone layer

There are many reasons for the hole in the ozone layer. The main reason for the hole is to let out waste products. Everything that lives and breathes has a hole to let out waste products, and the Earth is no different. The biggest product leaving the hole in the ozone layer every day is water. Pollution and other kinds of gases also go out the hole in the ozone layer and are swept out into the universe. The hole in the ozone layer is at the bottom of the Earth because life is moving upward in space. The ozone layer regulates sea level and temperature and keeps out harmful radiation from the sun. The tides have a lot to do with the hole in the ozone layer and have nothing to do with the Moon. As the Earth rotates, it is tilted on an axis of twenty-three degrees. The thrust of the sun is greatest on the Earth at the two poles. It swirls around the poles, pushing them towards each other, causing the Earth to bulge at the Equator. As the thrust of the sun swirls around the poles, it bends power from the sun around the poles, sometimes causing the northern and southern lights. The Arctic ice cap in the north is a floating ice cap and the Antarctic in the south is solid ice on solid rock. The Antarctic ice cap goes in and out of view of the hole in the ozone layer. When the Antarctic ice cap is in view of the hole in the ozone layer, the thrust from the sun coming through pushes the Earth up in the oceans, while at the same time holding the Arctic ice cap down in the north. This causes the waters to go down. When the Antarctic goes out of view of the hole in the ozone layer, the Earth goes back down, and the waters rise. This is what causes the tides to rise and fall every day. More pollution leaves the Earth through the hole in the ozone layer in the summer than the winter because in summer, the hole in the ozone layer is tilted more away from the sun, while in winter it is tilted more towards the sun. In winter, the cold and flu are more common than in the summer. In the summer the air is fresher. It is because there is more pressure acting on the hole in the ozone layer in winter than the summer that causes this. The prevailing winds come from the southwest. They are driven north by the thrust of the sun coming through the hole in the ozone layer and from the west because of the corkscrew effect of the Earth as it rotates into the east. The trade winds come from the east because the Earth rotates into the east. In winter, sometimes the north winds blow and bring snow. This is because as the Earth comes back down its orbit into winter, the broad end of the orbit, and the Arctic has tilted away from the sun and is backing into the dark night of the universe, receiving no heat or light from the sun. Pressure builds up behind the Arctic, and it gets very cold and sometimes causes the wind to blow south, and depending on the moisture in the air, we may get snow. We could not exist without an ozone layer, and every planet man settles on will have to have an ozone layer with a hole in the bottom.

-

The Earth is The Center of the Universe

Some people like to write a theory. I like to try an explain a fact. A theory is nothing more than a "what if." You can't build anything on a theory—only another theory. God the Father, the Almighty, created all that is seen and unseen. Some believe a little rock exploded billions of years ago and became everything. There was no mention of water in this theory. There is no rock without water, as we know. This theory has been around for some time until one day a little boy asked his father, "Where did the rock come from?" Somebody had to quickly write another theory, which said all the dust in the universe came to one point in space and became rock. Again, there was no mention of living water. Salt and water are the two building blocks of life. Both have existed from all eternity. God is salt, water, power, and light. God is infinite love, and love is the source of all creation. First God created Heaven, and then He created the Earth. God dwelled in the void from all eternity. God said when dwelling in the void that nothing consumed the universe all around. After God created Heaven, he created the Earth around the void. The structure of the void was a wall of living water in the shape of a heart, which is the symbol for love. A three-dimensional view of the void would look like a walnut. God the Father dwelled in the void with the eternal flame in the center of the void. After God created Heaven, he created a ball of salt around the void and beyond this, a wall of living water. God the Holy Spirit can change in substance the makeup of things, so God made salt and water out of love. Love is pure energy. So the Earth was built with a stellar core. So as all stars are made out of Saltpeter, the Earth to was built like a star. When God stepped out of the void to go up to Heaven, taking the eternal flame with Him, he ignited the Salt Peter, which burned and raged outward, leaving a vacuum behind Him at the center of the Earth. As the Salt Peter burned outward, gravity increased, and by the time the salt was completely burning, gravity pulled the wall of living water back down onto the salt, creating the crust of the Earth. When God stepped out of the void, he left his presence behind at what is now the center of the Earth, sealed in forever and ever. The Earth is the only place in the universe that has eternal existence. Heaven was only created over nine thousand years ago. Popes in Rome thought for 1,500 years with the gift of infallibility that the Earth is the center of the universe. Jesus Christ is the head of the Catholic Church and is never contradicted. The Earth is the eternal home of mankind. It is eternal because it is the only place that has eternal existence. That is why it is the center of the universe.

The Speed of Light

The speed of light is an incorrect statement. Light simply shines, and it only shines as far as it can reach, depending on the size of the fire. It is power or electricity, natural or man-made, that travels at a very high speed, 186 thousand miles per second. The sun produces power or energy at the very inner core of the sun, next to the black hole in the center, in the process called nuclear fission. In nuclear fission, under severe pressure and heat, the nucleus of the atom splits into many smaller nuclei, which become the nuclei of new atoms. At the moment the nucleus splits, energy or power is produced and loops into the black hole just a little and runs back out into space with gravity. Power is like a tiny little pipe or a space canal that gravity runs throughout into space, just like a vacuum runs through a pipe. Thousands of these tiny little space canals would fit in the palm of your hand. It is the gravity running down through the tiny little space canals which break up the power into seven different wavelengths of seven different colors. It is these seven different colors of power with electrons that give us daylight. Light illuminates that which it reflects off of, like the Earth and its atmosphere. If light were traveling at 186 thousand miles per second from all the stars in the sky on a clear night, it would be bright all night long. A star is a big ball of saltpeter burning as a raging fire in a vacuum in space. At close range to a star, thrust outweighs gravity, and everything is thrown off. Nothing is every pulled into a star.

The Wheel

The wheel came into existence almost by accident. The wheel is man's greatest invention. This is the story I heard. Many centuries ago, a little six-year-old girl named Linda Smart was going home from her neighbor's house one night, sitting on her father's shoulders. The Moon was up and it was full, and all the way home, Linda was looking up at it. Just before they got to their house, Linda asked her father if he would make her one of those "round things" in the sky. Sometime later, Linda's father was cutting wood and he remembered what Linda had asked him. He cut a slice off the log of the tree and gave it to Linda, and Linda went pushing it all around the field playing with it. Then a neighbor saw Linda playing with the wheel and got a bright idea. He asked Linda for the wheel, and she gave it to him freely, and he knocked a hole in the center and put a shaft through it. Then he mounted his barrow on top, and man was on the move. Linda's father saw this and cut more slices of the log for his children. Linda Smart gave her gift freely to the world. The wheel is man's greatest invention because of its simplicity. Simplicity is the height of excellence. God created all things the simple way by practicing the mechanics of love.

-

James Watt

James Watt, as a seven-year-old boy, was sitting in his mother's kitchen watching the kettle boiling when he envisioned the power of steam as he noticed the steam from the boiling water lift the lid of the kettle. This was the beginning of the industrial age. James Watt took man from behind a horse and put him on the face of the Moon. A six-year-old girl and seven-year-old boy gave their gifts freely to the world. These gifts came out of innocence— not greed.
-

The Human Person

A human being is of three parts: heart, soul and body. The human heart is the electrical grid, or ghost of the person, from head to toe under the skin. The ghost of the person is produced by the human soul, which is in the left side of the organ called the heart. It is in the human soul where power is produced, which powers the human heart and pumps blood around the body, giving life to the flesh. It is the spirit that gives life; the flesh is of no avail. In the center of the human soul is a tiny little star which produces power like every other star in the universe. Every star in the universe has a pulse. The smaller the star, the faster the pulse; the bigger the star, the slower the pulse. The pulse of our sun is approximately eleven to twelve years. A little baby's pulse beats a lot faster than its parents. It is the pulse of the star that makes the heart beat in our chest and also pumps blood throughout our body. The human soul is the Firestone in the Bridgestone of the human heart. The human body is the most brilliant machine ever built. At the beginning of time, people lived for many centuries, not decades. The human body is mostly muscle and bone, both of which start at the feet. The left side of every organ in the body is bigger than the right. The left side of the face, the left hand, the left foot—everything on the left is bigger than the right. There is only one contradiction to this, and that is the thing inside your skull, your brain. It faces downward, and the right side is bigger than the left. The unique individual resides in the soul—not in the head. The human heart (the human ghost) is full of intelligence from head to toe. You might wonder what's in your head. Are we wrong about something? People have been blowing stuff down their nose from their head all their life. In early times, some people drilled holes in their skull to let out demons. The truth comes from the heart, and lies come from the head. A doctor in the U.S. removed half a brain from a person's head and sometime later they went back to have a look, and it had grown completely back. Fingers and toes don't grow back. An ear or a nose doesn't grow back. If you pull a weed out of the ground and you don't get the roots, it will grow back stronger than ever. Some Christians put a bumper sticker of a fish on their vehicle as a symbol of Jesus Christ. The salmon of knowledge does not have a brain in its head or a chip in its tail, and neither did Jesus Christ. He did not receive the seed from Adam, which was stained with original sin. The stain of original sin can be removed, but the alteration has remained. Somebody set up shop in Adam's head when he committed original sin. The nose on your face is there to split the resistance in front of you off to the two cheekbones so that you can move forward without getting a headache. If you had a flat face and you wanted to ride a bicycle, you would get a massive headache. The two soles on our feet are meant for walking on. When we pound down on the heels of our foot, and stone meets bone, we are creating stress. All the bones in the body are connected, and this stress goes up through the whole skeleton. If man could learn how to walk properly and land on the two soles of our feet, we would be using our muscles instead of our bones. The soles are a pad of muscle, and since all the muscles in the body are connected, we would be toning up our body as we walk and keeping ourselves in good physical condition. The organs are also muscle. As we walk, if we toe the front of our feet inward slightly like the front wheels of a car, it causes the front of the foot to tilt downward a little. As you barely touch the ground with your heel, it causes the front of the foot to tilt down more, transferring most of your weight onto the sole of your foot. This would be the same for the next foot. Toe in, tilt down, and tone up.

Progress

We have no progress in the world today. What we do have is a world in a state of stagnation swimming around in a sea of iniquity. America was once called the land of progress. It has been almost fifty years since man landed on the Moon, and not much good has been done since. All stagnation is good for is being a breeding ground for death, disease, and disaster. That is what we are all looking at today. The Pope in Rome over forty years ago warned about the attack on the family and the consequences of it. We see the family as being pulled in all directions, and now society is near to collapse. Pope Benedict the XVI said the western world is in the same position as the Roman Empire was at the point of its collapse. What has happened in the world is that just about every wrong we can think of has been appeased. And then these things become the acceptable norm of the day. If we could imagine we were in a land of progress, it would bring prosperity and peace to the people. Progress is like a tonic for people. It lifts up their spirits when they get involved in doing good works. Examples of progress are going from the candle to the light bulb, indoor plumbing, the wheel— the greatest advance of all—, and the engine, the plane, and so on. We have to change from a culture of death to a culture of life, or we can't spread life across the stars. Soon man will go to Venus and Mars and bring these two worlds to life. This is going to take a big effort by lots of people. We are going to have to build a new transportation system to get to other planets and make new ways of doing things to move us along on the road of progress. We have to get progress on the move again. We cannot just sit here and do nothing, spitting and shouting and throwing bombs at each other. We've tried this for nine thousand years, and it hasn't gotten us anywhere. What we need to do is sit down with humble hearts and simple minds and start to figure out our way out of this mess.

The Earth Is the Oldest Body in The Universe

The Earth is the oldest body in the universe. God created Heaven before he created the Earth, but Heaven has no bricks and mortar to it yet. After God created Heaven from the void which he dwelled in from all eternity, he created the Earth. The Earth was created with a stellar core and a wall of living water outside it. When God stepped out of the void and went up to Heaven, taking the eternal flame with him, he ignited the saltpeter, which raged outward, and then gravity pulled the living water down on top of it. When God came back down from Heaven to create all kinds of life, he lit the two first stars in the sky. He lit the higher light to rule the night and the lower light to rule the day. Our sun is the only place that could rule the day because there was nowhere else. When the sun was lit, the Earth would've been at the farthest point from it on June 21st. Then the gravity of the sun would've pulled the Earth back down towards the sun, and the thrust of the sun would've pushed it away. Here on Earth, when we cut a tree down, we can find rings in it for year of growth so we can tell the age of a tree. We had a flood here seven thousand years ago so it would be hard to find rings anywhere to tell us the age of the Earth. There is one place on the Earth that didn't turn upside down during the flood and that is the Arctic ice cap, which is a floating ice cap. In wintertime, the Arctic freezes from the outside in, pushing dead air in with it, and when it gets toward the center it pushes it down. You may have noticed the blackness on the underside of a sheet of ice, if you ever walked on a sheet of ice. Then in the summertime when the temperature in the ice rises, the black blob under the ice stretches out, forming a ring. If I were a betting man, I would say there are today (2018) 9,022 rings under the Arctic ice cap after June 21st.

Immorality

Immorality is a useless waste of valuable time. Immorality destroys faith, hope, and trust in God and each other. Today we're living in the age known as the culture of death. We are also living through the great age of appeasement, where we have shown indifference to just about every wrong in the world. We are living today in the worst time in all of human history when tens of thousands of little children in the womb are being slaughtered every day around the world. Here in the U.S., we slaughter on average 3,500 children a day in human slaughterhouses. I don't have the numbers for every nation, but approximately one hundred thousand children in the womb are slaughtered each day around the world. In the world today we have infanticide, euthanasia, abortion pills, and some women have devices implanted in their body that also kill newly conceived children. If we are not fruitful and multiply, we cannot go to new worlds and build new homes. There is very little hope in the world today. Most people don't trust many other people. The reason we are in this mess today is because good decent people sat back, did nothing, and said nothing in the face of terrible ignorance. This has allowed evil to crawl all over this world. The biggest problem in the world is the truth is not being spoken very much. The truth is the only weapon to confront evil and expose it for what it truly is. When we see evil raise its ugly head, it should be cut down with the truth. The truth does not offend or insult anyone; it is simply the truth. It is a person's own ignorance that offends them when they hear the truth because they are guilty of wrong. All truth is not good, but it is all fact. A great lack of hope in the world causes a lot of depression. We can't expect our spiritual leaders to do everything. Politicians and governments have failed the people. Most governments around the world are corrupt, so we the people will have to start determining our own destiny. We cannot just sit around here idle, going nowhere. You cannot go against the grain of life and expect to succeed. It would be terrible to take, take, and take all your life and lose your soul to the abyss, which is the bottomless universe. There is no wrong in life that cannot be rectified by God. If we could give hope to people in doing good works, and if we could start defending the truth, especially when others speak words of hate, and we could stop omitting the existence of God who created all things, then those who could, should inspire the good in the hearts of men. Words are not enough—we need to take action. We are going to have to start doing things in simpler ways to progress. Progress demands that we simplify the ways of doing things. Like crossing the Atlantic in seven hours instead of seven weeks. Before there was a wheel, there was no wheel. We can take giant leaps like this again and again if we simplify our minds and humble our hearts in the face of Almighty God, who knows, sees, and understands all things. Our time down here under Heaven is brief compared to the rest of eternity. Please do not throw it all away working for Satan.

Eternal Existence

For everything in life, there is a source. The source of all things is love. God is infinite love. Love is not just a word; it is something real. It is pure energy, pure goodness, and kindness. Love is the raw material God used to create all things. For everything in life that was made, effort had to be made. Nothing ever did or ever will happen without effort. All effort in life, physical or spiritual, is mechanical. There is no such thing as scientific effort. For everything that was ever made, there had to be a maker, the Creator with a pair of hands through which God created all things. God the Father, the Creator in the flesh from all eternity, is the instrument through which God created all things. For everything in life that was made, there had to be a maker with intelligence to reason, as we can see order and harmony to things in the night sky. All life comes from the heart of the Father, and the identity of every human being was in the heart of the Father from all eternity, each known to the Almighty. If there was no eternal existence in the universe, there could be no existence at all. The Catholic Church introduced modern science to the world five hundred years ago to help explain some of the simple facts of life around us. Today, science has gone haywire. We have a mountain of theories that explain nothing and nobody understands them, scientists are ten a penny, and I think it's time to throw the theories in the trash and start explaining some of the simple facts of life around us.

Man is Blind

God left man blind because of sin. We do not have perfect clarity or complete peace in our hearts, which makes it hard to see and understand the simple things in life around us. The more sin we commit, the blinder we become. Mortal sin turns the Firestone of the soul red when it should ideally burn white so that we can walk in the light as Jesus did. As time has passed, man has gotten more and more confused. We tend to do things in complicated ways, and we expect everything in the universe to be complicated when it is not. It is the simplicity of God's creation that has eluted man since the beginning of time. The mechanics of the universe are no more complicated than the workings of a wheelbarrow. Some people think they have a higher education and understanding of things. There is no such thing as being highly educated, and there is no such thing as having a higher understanding. Understanding is always simple when you get it, and understanding is only found in the simple truth. All truth is simple—good or bad. Some truths are so simple we cannot see them. One day Jesus looked up to his Father and said: "You gave them eyes, but they cannot see." Another day, Jesus said to the people "Yee are the blind being led by the blind," which means every one of us is blind. There are no experts in the universe but one, and that is the Almighty Himself, who knows, sees, and understands everything. Our time under Heaven is about service. We are all down here to serve—not to be served. When Jesus came down, he said, "I came down to serve, not to be served." It was when I remembered that Jesus told me I was blind and when I admitted my blindness that I finally started to find a little more and more clarity. With a simple mind, a humble heart, and a will to serve God and our fellow man, we can then start to make a little progress. When we do get the wheels of progress on the move, we will not let them be halted by ignorance ever again.

Pollution

There is a lot of talk about climate change today. The weather changes from day to day and from season to season. There are more seasons than the four seasons of the Earth as it goes around the sun. There are the four seasons of the Milky Way as it rotates. Sometimes we may be at the front or the back of the galaxy, and sometimes we may be facing in towards the center of Excelsior or facing away from the center. There are also the four seasons of Excelsior. We don't know where we are and what we are doing, so then why are we complaining about something we know little about? It is true that the air and the waters are polluted. Most of the air pollution is going down and out of the hole in the ozone layer, but there is still a certain amount between the ground and the ozone layer. If we're going to use solar power, then we need to use it properly. We're going to have to a build powerplant to harness the power of the sun. There is a greater abundance of power up above the ozone layer that is not being used. If we tap into it properly, it will serve all our energy needs. This would eliminate the need for all batteries and cut down on air pollution. The car will be powered fuel-free, but we have to do a little more for the car. We have to make the car crash-proof. Different-sized receivers are needed for different applications. The car hasn't made a lot of advances in the last hundred years. With a receiver in a car, with plenty of power which we can turn into electricity to power the car, we can also build a force field around the car to prevent an oncoming impact. This force field could also be used to do the majority of breaking, lengthening the life of brake pads. Pure power coming down out of the sky is harmless until we turn it into electricity, which can be dangerous. This is one way we could eliminate all air pollution.

The waters are also polluted. Effluent from human waste is flowing into the oceans. Some of the poorer parts of the world don't have sewer systems or proper sanitation, and human waste is flowing into the seas. One night I watched a show on television about a major river where people were taking a bath, and human waste was flowing past them down the river. This was one of the biggest rivers in the world.

When these waters go down deep in the ocean and are pressurized in through veins in the Earth, no one knows where they rise. We are all drinking from the one well. Filtration does not get rid of the harmful pollutants in human waste—they are invisible. A certain amount is filtered as it goes through the Earth, but a certain amount comes back up as rivers and streams and into our reservoirs and wells. Distilling polluted waters would kill the harmful pollutants and leave the waters safe to go back into the ocean, or we could go to the source of the pollution itself, the toilet, and see what we could do there.

Noise pollution is a problem nobody is talking about. We are bombarded with noise pollution from jet engines flying all over the world. Noise vibrates the air and creates stress. Gravity pulls this down into the Earth through everything on its way. This can't be good for us. Things that make noise here on the Earth, like jack hammers or rigs for drilling wells, need a way of absorbing the loud noise. Snow absorbs noise and human flesh absorbs noise, but these things are not practical. We have to look for other materials that will absorb noise and place them at the source of noise.

Venus and Mars

Soon man will go to Venus and Mars, but we need to build the transportation system to get there. One evening I went for a walk around the block, and when I got to the corner of the block, there was a large puddle on the side of the road, and there was an empty shoebox on one side. A gentle summer breeze came that evening, and I watched it sweep the box across the still water with the greatest of ease. In my opinion, the canal is the greatest piece of construction on the face of the Earth. The canal allowed man to transport large quantities of goods over great distances with the greatest of ease on still water. So then I thought, could we build a canal through space, a path of no resistance, to safely travel through with the greatest of ease? A spaceship could be a machine of four or five decks that would carry approximately five hundred people comfortably. A spacecraft will have an independent field of gravity of one and will also have shields around the ship. The shields around the ship are the atmosphere that the gravity of the ship takes with it as it ascends up into the sky. Power from the ship can be used as a protective shield in this atmosphere around the ship. Once we build the spacecraft, which can achieve unlimited speed—many times faster than lightning—we have a way of getting to and from planets in a short period of time. The shields around the ship look like the shield of life, or the void God once dwelled in.

Building a spacecraft is going to take a lot of manpower, and they can be built on assembly lines. We are not going to Venus or Mars with one or two ships; we are going to send a fleet of five hundred to each planet on the same day so that people from all nations who wish to be part of this event can be represented. We are not going to Mars or Venus to pick a few rocks and bring them home. We are going to bring these two new worlds to life and build two new homes. The first thing we have to do is build the oceans and the atmosphere on both planets. It's true Venus is upside down, but a thousand ships would have no problem putting her back on her feet. Venus got too top heavy and had no oceans to support it upright. There is a good probability that Venus had two moons around it before it turned upside down because there are two moons in our solar system in places they don't belong. There is one on the asteroid belt, but moons don't grow around asteroids, only around planets. There is also a moon going around one of the outer planets, rotating in the wrong direction—it did not grow there. As we build the oceans and the sky above, an ozone layer will start to grow in the atmosphere. The swirling thrust of the sun will form a ring above the south pole of the planet, and from there it will stretch outward around, up and over and in over the top of the planet, so we do not have to worry about ozone layers. An atmosphere is mostly water and air. The Earth produces an abundance of water every day, and it was water from here that allowed other planets and rocks to grow in the universe. Once we have the oceans and the sky built, the ozone layer will regulate the sea level and temperature beneath, and then we can install a powerplant. With a powerplant installed, we can bring machinery powered by receivers to these new worlds. On the morning man lands on Venus and Mars not long from now, we'll lay down the foundations for two new capital cities. These two cities will be sister cities founded on the same day in history. These two cities will have an oval-shaped park at the center with an inner row of towers reaching three hundred floors into the sky and then tapering back down and out to the suburbs. We will set city limits and not let cities sprawl all over the land.

We will start at three hundred by using gravity to our benefit, and more and more modern materials—new steel. When we use gravity to our benefit in the proper manner, it will allow us to go higher and higher and still give stability to the towers as they go higher in the sky. New steel is also going to be a great benefit to us as it will be many times stronger than the steel used today. The capital of Venus will be called St. Helen's, and the capital of Mars will be called Destiny. St. Helen was a great lady and named all the holy sites for us in the Holy Land, and Venus is supposed to be a woman's world. The capital of Mars in called Destiny because it is man's destiny to go west and settle the stars. Planets have to be timed and set to a 24-hour day and a 365-day year, the same as the Earth. All times and all seasons obey God's law. We will also have to set these planets to a gravity of one. With the oceans and atmosphere around Venus, we should be pretty close; we may have to adjust a little. The oceans and atmosphere around Mars won't be enough to bring gravity up to one. So, we will have to add more material above the ozone layer until we get to a gravity of one. The more mass that has to be pulled in towards the center of a planet, the higher the gravity. There are oceans and oceans of gases out there in the skies all around; we just need some heavy, clear odorless gases to add into the upper atmosphere. Imagine you have five pounds at the end of a rope. The pulling effort needed is not that great. Add another fifty pounds, and the pulling effort required is a lot more. With a powerplant installed in the center of each park, the first things we need to build are the sewer and water systems. With more modern ways, these things won't take long. The surface of Venus is covered in a lot of ash. When Venus turned upside down and was rotating the wrong way, thrust got weaker and weaker until the gravity of Venus pulled the lunar orbit back down onto Venus, and it is also receiving dust and ashes every day. The surface of Mars is covered with stones, which means it's receiving a lot of water. Stones can't grow without water. We are going to have to prepare tracks of land for farming on Mars. So we have to get rid of the stones. If we attach a device on the front of spacecraft, we can project a band of sound, slightly lower than the sound of silence, down onto the ground, and it will crush the stones into dust, and we can also penetrate the ground a foot or so. A lot of farm machinery pick up stones in the fields and sometimes cause breakdowns. The stone crusher would eliminate this. As we sweep the fields of Mars with stone crushers on the front of spacecraft, we can also harrow the ground from the back of the ship with the tractor beam.

This will aerate the ground for us. Next, we dress the ground with cattle slurry. Into the slurry, we add grass seed and earthworms. Earthworms will aerate the ground for us and help the grass to grow. Then we need to make it rain to wash in the slurry, and next we apply fertilizer. Farmers would have tested the soils beforehand so they'll know which fertilizers to apply. Then we need more rain, and then finally a coat of peat moss, and then more rain. The grass seed should be up in a few weeks, so we'll have green as well as blue and white. We will also need to plant thousands of acres of trees—trees that produce the most oxygen first. This will take many years for trees to be fully grown, so we are going to have to keep oxygenating the air until such time. The tree allows man to live and breathe. We do not know yet where rivers are going to rise and which way they are going to run. We do know that here on Earth most major rivers rise in the north and flow southward. The people of Mars, Venus, or any other new worlds will determine their own destiny. We here on Earth can give them a start.

Manpower

Big projects take a lot of manpower. Building new cities will take hundreds of thousands of people from different trades. We can use spacecraft to prepare ground for the farmers to make land. There is no land on Venus or Mars. Land has to be made, and it's farmers who make land. They will be some of the first people who go to these new worlds. The farms of the future will be a lot bigger than the farms of today. We here on Earth have a population of over seven billion, so we are limited to how much we can do. We are slaughtering millions of innocent children in the womb all the time. Here in the US, we slaughter 3,500 innocent children every day in abortion mills. Not only does every human being have the God-given right to life from the moment of conception, but every pair of hands we receive we could do something with right now. Abortion is the most horrific crime ever committed. It is the most cowardly act ever committed. It is as far from bravery as one end of the universe is from the other. Not only are we slaughtering the help on the way into life, we're also losing thirty percent at death to hell and damnation. Right now in the year 2018 we are in the worst state since the beginning of time. Some people wonder if there is alien life out there. The only alien life out there is those we alienated. One hundred years prior to judgement day, which is just around the corner, we will have slaughtered four billion innocent children in the womb. Jesus spoke about these people before he went back up to Heaven. He said,

"Whatsoever you do unto the least of my brethren you do unto me." These are the very least of his brethren, the littlest of human beings. When Pope John Paul the Great spoke about those UFO's in the sky, he said "Do not be afraid of these people." These people are the ones we aborted back up into the face of God. They do not mean us any harm, but they will patrol the universal highways and byways of space for the next thousand years. So where can we get manpower? We could open our hearts and receive graciously every blessing God sends us. A newborn baby is *not* a burden on society, and abortion is *not* a necessary evil. These are words of hate. Can you see our stupidity, our ignorance, and our greed? This is what happens when people remain silent in the face of terrible ignorance. It is the silence of appeasement that has allowed evil to crawl all over this world. Soon the dead will be raised, and some will be back here among us on the Earth, but if we don't start growing in numbers, we can't spread out very far. We need to be promoters of life to spread life.

Judgment Day

Judgment Day is not very far away. Pope Francis has declared a year of mercy in the Catholic Church, and he says, "Mercy is there to be shown to all." Jesus said "Those who do not go through my door of mercy will have to go through my door of justice." Jesus is coming in glory to judge the living and the dead. The saints in Heaven don't need to be judged, and those in Hell and damnation don't either. None are ever going to be raised out of Hell and damnation. At death, Jesus calls every soul before him and makes clear who and what he is and their only hope of reaching eternal happiness. It is up to each individual then in that moment of mercy to accept Jesus completely or not at all. Those who rejected Jesus are in Hell and damnation. It is the good and the true who accepted Jesus who are in the dust, which we call Purgatory. The people in Purgatory are doing penance and purifying their hearts. They are all guaranteed to get to Heaven, unlike us, here in the land of the living, who are not. Judgement day will be our final moment of mercy. On judgment day, Jesus will take away death from before all who desire in their hearts to reach eternal happiness. Disease, hunger and poverty will be done away with. The divine physician is coming with the cure for all that ails. On judgement day, before Jesus comes down, three billion saints who are at rest in the sacred heart of Jesus will go marching into Heaven. There are nine and a half billion lost to Hell and damnation who will never see the light of day again. And then there are fifty billion practicing Catholics in purgatory to be raised back up into their mortal bodies. Three billion of them will be back here on the Earth, and the other forty-seven billion will be raised up into the halls of paradise. Paradise is located on the asteroid belt, and it looks like a golden pear, speckled green, blue, and white, or like a light bulb upside down with a wooden stair coming down from the bottom. No one will be harmed on judgment day. God never sent any man or woman to Hell. All who got there got there on their own two feet and of their own misused free will. Free will is a gift from God to choose from all that is good in life. Money will go out of circulation on judgement day, which will bring a lot more freedom to people everywhere. Mary, our Blessed Mother, once said "Oh, how the mighty will fall." This will sort out the men from the mice. If we do not change our ways for the good on judgement day or before and start doing the will of God, which is the will of man, eternal death is still going to be available after judgement day. Mortal sin is the most dangerous thing in the universe. It is deadly—worse than nuclear waste—and has to be disposed of. It has to be deposited into Hell or to the core of the Earth for now. On judgement day or very soon after, people will have to decide whether they want to choose life or death. Hell and damnation are final. Jesus won't be taking a whole lot of people from here up to Heaven on judgement day because there aren't that many practicing Catholic saints walking the Earth. You also need the bread of life for the journey home to Heaven. Judgement day will be a breath of fresh air for most people. It's the gluttons of history who are in Hell and damnation, and we can't afford to lose anymore to Satan. People need to be careful in what they are doing and start living the right way if they wish to escape Hell and damnation. God is not mocked.

Leadership

America is chosen by God to lead mankind out of the womb of creation and land on two new worlds at this time in human history. Before we go anywhere, mechanics have to be put to the test. Mechanics are always well tested and approved before they are given out for public use. Here in the US, we will build a test shuttle and test out all the mechanics on the spacecraft. This test shuttle will be given the name The Simple Way after Mother Teresa of Calcutta, who preached all her life that life is the simple way. Then after this shuttle is tested and approved, we can start building fleets of space ships. America is blessed more than any other nation in history. People from all over the world came to these shores seeking a better life for themselves and their children. They wanted to live in freedom and have the right to practice their religious beliefs without being dictated by warmongers and tyrants. At this time in the US (2018) we have a chance to do the right thing and be a good example for others to follow. There are plenty here in America and around the world who are ready for a good challenge. There are some here in America, domestic enemies, who are trying to divide the people against each other, and there are some here in America who are gnawing at the foundation of this nation. They might as well be chewing on pig iron. The foundation of this nation came down from Heaven in the words "We the people hold these truths to be self-evident that all men are created equal endowed by their Creator with certain inalienable rights, life, liberty, and the pursuit of happiness." Jesus Christ is the eternal word, and He is the foundation of this nation. It is in humble confidence that anyone can give leadership in the right direction. I remember when Pope Benedict became Pope, he addressed the people in these words: "I am your humble servant. Humility is our greatest asset."

The Rosary Beads

The rosary beads are a set of beads that Catholics use to pray the rosary. With the beads in a circle and the Cross pulled tight, you are looking at a blueprint of the life. Where the necklace comes together is the place called Heaven, and the cross at the top is the place called Eternal Glory. All the beads on the necklace are also part of the Kingdom of Heaven. The stars are here at the center of the beads, one billionth the size of a grain of salt. Man cannot reach up to God, but God can reach down from Heaven and lift us up. Jesus compared Heaven to a mustard seed that grows into a very big tree. Heaven is the seed of the Tree of Life with the cross at the top moving forever upward. The kingdom of God, which is inside the necklace where we are serving our time, is the root-ball for the Tree of Life. As we build new rings of fire with the help of God in times to come, we will see the Tree of Life starting to grow. I have seen the roots of the Tree of Life coming down from Heaven into the kingdom of God. They are the flow of saints going up to heaven from the kingdom of God. As man builds new rings of fire at the center of the rosary beads with the help of God, we will be designing the figure of Our Lady. There is no end to how much we can build. As the kingdom of God grows so too will the kingdom of Heaven grow. Jesus said, "In my father's there are many mansions." The cross is the way of life which moves forever upward at infinite speed. Out beyond the rosary beads, God is a shield onto those who put their trust in Him.

The shield that God has built is a wall of living water. Jesus Christ, who is universal gravity, causes power, who is the Almighty, to run straight up through the living water, and universal gravity pulls down, giving us the heart-shaped figure of four chambers of life. We are not bound by space or time, and the Tree of Life will grow forever and ever and also the kingdom of God. Man is infinite in numbers.

The Seven Kinds of Life

Seven is a popular number in life, and there are many examples of this. The seven kinds of life are Mineral life, Vegetable life, Animal life, Human life, Everlasting life, Divine life, and Eternal life. Mineral life has existed from all eternity. An example of this is living water. Vegetable life, which God has created, is all that grows on the Earth. Animal life, not to be confused with Human life, seeing as animals don't have a soul, was also created by God. The male of each animal carries the life of the young in him. Human life is created by God at the moment of conception—heart and soul. The earthly parents bring the ingredients for the flesh. That is why we are called procreators. Everlasting life is a share from eternal life, which we can receive at Holy Communion, but we are not perfect down here, so we need to keep receiving it until we go to Heaven. Divine life is the three perfect and most pure natures of the Blessed Trinity in the center of the Eternal Flame. This is also called the Holy Trinity—One in Being. Eternal life is Jesus Christ. He is the Eternal Flame. He is the Life, the Truth, and the Way. Through Him all things were made, and from Him all good things come and not by our efforts alone.

It Takes Seven Persons to Create a Human Being

It takes seven persons to create a human being. God the Father, the Son, and the Holy Spirit, Mary our Heavenly Mother, two earthly parents, and the one being created. God the Father the Almighty takes love from Mary's Immaculate heart, which is changed in substance by the Holy Spirit into a tiny piece of saltpeter, which is ignited through the Eternal Flame, Jesus Christ. At this moment, the Almighty takes love from the Father's heart with the identity of the one being created, which is infused into the firestone of the soul. The firestone of the soul is a tiny piece of saltpeter one billionth the size of a grain of salt. At this moment every human being has received all their God-given rights and their gift of free will. Then, by the power of Holy Spirit, the individual is taken down and placed in the mother's egg during sexual intercourse, which is an invitation to God to create a human being. The firestone of the soul encases itself inside a wall of quartz and then radiates power out through the quartz into the electrical grid of the human ghost, or human heart. At this time, the individual inside the egg gives a signal for the father's sperm to come and fertilize the egg. Men have both male and female seed. If there is male inside the egg, a male seed is allowed to enter, and if it's a female, then a female seed. The seed cannot force its way through the egg. It is at this moment of fertilization that we were all stained with original sin.

There are three types of sin: original sin, venial sin, and mortal sin. We can have the stain of original sin removed in baptism. Venial sin is not a serious sin, but if it mounts up, it can become serious. Venial sin is the dust being thrown off at the outer layers of firestone of the soul. There are seven types of mortal sins, also known as deadly sins, and pride is the deadliest of all mortal sins. Mortal sin is committed at the inner stone of the firestone where the individual decides to do serious wrong and throws out heavy metal like the black spots you might see on the sun. Mortal sin is very dangerous and has to be kept in a safe place inside the wall of our soul or in Hell or damnation. Damnation is the bottomless pit where Satan is right now for the next one thousand years. We need to be careful and know what we are doing and not get caught at death with mortal sin on our soul. It is true that Jesus is there with mercy to be shown to all who have manners to ask for.

The firestone of the soul is a pulsating star just like every star in the universe pulsates. The baby has a faster pulse than its mother, and its mother's is faster than its father's pulse. Our sun pulsates every eleven or twelve years. Some other bigger stars will pulsate maybe every fifty or hundred years. We can feel our pulse at the end of our wrist just above our hand. The human soul is located on the left side of the organ we call the heart in our chest. It is our soul that makes our heartbeat and draws our breath. Blood is produced with salt and water and power that comes out of the soul. Ideally we should keep our soul burning white and not red like crimson. You can't see too clearly walking around in red light. We need to keep our soul maintained and working properly.

When I was a young boy, my father taught me to close my right fist, strike my heart, and say the words of St. Thomas under my breath, "My Lord and my God" at Mass when the communion was held up. Here in America people put their right hand over their heart when the national anthem is sung and when people take an oath.

Man is the Inventor

Man has been inventing things since the beginning of time—all kinds of tools and household items like the churn, for example— but the wheel is the greatest of all time. It was a simple innocent wish, almost by accident, but the industrial revolution has brought many new ways of doing things into existence. We have the train, the plane, the car, and we have radio and television, and today we have the computer and cell phones. We have washing machines and stoves and indoor plumbing. We have power and light. Thomas Edison, who gave us the light, was probably the greatest inventor of the twentieth century, and if you wonder where he got the lightbulb from, maybe he got a glimpse of Paradise.

    I see great possibilities the future, if we start to learn how to practice the mechanics of love—there is no end to what we can make. New steel, many times stronger and lighter than present steel today, copper, diamonds, and many other materials that will be new to us. Maybe the best way to bring something new into existence is to identify a need and figure out how to make it possible. We don't need to be digging in the dirt for all kinds of materials; we have plenty of materials up here to make all that we need. Salt and water are the two building blocks of life, and all we have to do is add a little TLC. We can all produce love, which is human life, but human life can also be hate, which is no good. The possibilities when we start practicing the mechanics of love are not just many—they are infinite. There are people making diamonds today, but they are not perfect. An imperfect one is not good to anyone. We will learn how to make perfect diamonds by the ton. There will be great need for diamonds in the future; we will be able to pave the roads with them, and we will never have to plow the snow and ice off the roads. Many other uses for diamonds in industry and farm machinery exist. This is just one example of what we can make. One thing we don't need to make is gold. Gold is a precious metal because it is clean. Gold is produced by the stars and thrown out by the thrust of the stars onto the orbits and is rolled by the solar winds. There is gold dust on the orbits all around the stars. Here on Earth, the surface of the planet is seventy percent water, and seventy percent of the gold dust lands in the oceans. Some of this gold dust is pressurized in through veins in the Earth deep in the ocean, and some of it comes back up as the source of rivers and streams. So we can harvest the gold on the orbits as much as we need.

In the future, we are going to have to build planets and moons, and then the time will come when man will feel the need to light a new ring of fire, like Excelsior. When we humble our hearts, and with God at our side, it won't be very difficult to figure it out. A star seed is about the side of one grain of wheat, which is the ingredient for one star. Billions of these stars will have to be put in a capsule the size of a beer barrel, which would be the ingredients for one galaxy. Then we need billions of galaxies and form an oval-shaped track, and we drop all our cargo and pull west, and watch a new ring of fire light up on the endless blackboard of the universe.

The End of Time

Some people are talking right now about the end of time, when what they mean is it's the end of the age. Others are talking about the end of the world, but this world has no end. Just a new earth with more land and less sea. God created time when he lit our star to rotate and the Earth, and it will end ten thousand years from that day, which will be in the year of our Lord 3004 on June twenty-first. Nobody knows when Jesus Christ is coming down to judge. But we can be sure that Satan and all his lot and all the lost souls will be taken down and out of life by the Father before the end of time. So then we can know that that day that we will all be in God, and God will be in all.

    I have seen a lot of great things coming in the future. Life is for everyone to take part in. Give it your best. Remember that humility is our greatest asset. Nobody goes to the Father who's not humble.

    Everlasting Life

www.ingramcontent.com/pod-product-compliance
Lightning Source LLC
Chambersburg PA
CBHW061519180526
45171CB00001B/252